The Science of Living Things

What is Hibernation?

John Crossingham and Bobbie Kalman

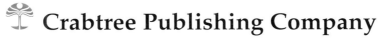

Crabtree Publishing Company

www.crabtreebooks.com

The Science of Living Things Series
A Bobbie Kalman Book

Dedicated by John Crossingham
For Julie Booth — because she hates pants

Editor-in-Chief
Bobbie Kalman

Writing team
John Crossingham
Bobbie Kalman

Editors
Amanda Bishop
Kathryn Smithyman
Niki Walker

Cover design
Kymberley McKee Murphy

Computer design
Margaret Amy Salter

Production coordinator
Heather Fitzpatrick

Photo researcher
Heather Fitzpatrick

Consultant
Patricia Loesche, Ph.D., Animal Behavior
Program, Department of Psychology,
University of Washington

Photographs
Frank S. Balthis: page 18
Stephen Dalton/NHPA: front cover
Wolfgang Kaehler: pages 8, 29 (top)
Robert McCaw: title page, pages 10, 14, 16, 19 (top), 30
Photo Researchers Inc.: Joe B. Blossom: page 5;
   J.L. Lepore: page 11
Tom Stack and Associates: Jeff Foott: pages 4, 19 (bottom);
   Thomas Kitchin: page 12;
   Tom & Therisa Stack: page 27
Other images by Adobe Image Library, Corbis, Digital Stock
and Digital Vision

Illustrations
Barbara Bedell: back cover, pages 4, 6, 8, 9, 13, 14, 25,
   26 (bottom right), 30
Margaret Amy Salter: page 19
Bonna Rouse: pages 5, 10, 11, 16, 17, 20, 21, 22, 23,
   26 (top, botom left), 27, 28
Tiffany Wybouw: pages 18, 29

Crabtree Publishing Company

www.crabtreebooks.com   1-800-387-7650

Copyright © 2002 CRABTREE PUBLISHING COMPANY.
All rights reserved. No part of this publication may be reproduced,
stored in a retrieval system or be transmitted in any form or by
any means, electronic, mechanical, photocopying, recording, or
otherwise, without the prior written permission of Crabtree
Publishing Company. In Canada: We acknowledge the financial
support of the Government of Canada through the Canada Book
Fund for our publishing activities.

Printed in Canada/112018/MA20181015

Library of Congress Cataloging in Publication Data
Crossingham, John
   What is hibernation? / John Crossingham & Bobbie Kalman
   p. cm. - (The science of living things.)
   Includes index.
   Describes the process of hibernation and the various ways in which different
animals use this process to survive in harsh climates.
   ISBN 0-86505-987-X (RLB) ISBN 0-86505-964-0 (pbk.)

1. Hibernation—Juvenile literature. [1. Hibernation.]
I. Kalman, Bobbie. II. Title. III. Series.

QL49.K294 2000   591.56'5--dc21   2001037209

Published in Canada
Crabtree Publishing
616 Welland Ave.
St. Catharines, Ontario
L2M 5V6

Published in the United States
Crabtree Publishing
PMB 59051
350 Fifth Avenue, 59th Floor
New York, New York 10118

Published in the United Kingdom
Crabtree Publishing
Maritime House
Basin Road North, Hove
BN41 1WR

Published in Australia
Crabtree Publishing
3 Charles Street
Coburg North
VIC, 3058

Contents

 # What is hibernation?

Many areas of the world have cold winters. In these places, snow covers the ground, ponds freeze, and food is hard to find. Animals that cannot live and find food during this cold season must find other ways to survive. Some travel, or **migrate**, to warmer places. Others **hibernate**, or go into a sleep state.

Hibernation is not like regular sleep. A hibernating animal's breathing and heartbeat become very slow, and its body becomes cold. The animal appears to be dead, but it is still alive—barely. Its body slows down so much that it needs almost no food. It gets all of its energy from the fat stored in its body.

*Many hibernators sleep underground, where they are protected from the cold weather. Underground chambers are called **burrows** or **dens**.*

True or not?

Small animals such as dormice, shown left, and Arctic **ground squirrels**, are **true hibernators**. True hibernators sleep for several months. During this time, their body temperature drops to just above freezing. Some larger animals, such as raccoons and bears, spend much of the winter sleeping, but they wake up often. Their bodies stay much warmer than those of true hibernators.

When is the right time?

Animals must begin hibernating at the right time. If they fall asleep too early, they may run out of body fat before winter ends and die. Scientists believe that the length of daylight tells most animals when they should hibernate. As winter approaches, there is less and less daylight each day. Animals know it is time to hibernate when the days reach a certain length.

Different types

There are many types of hibernation. Each type is suited to the animal's body and its **habitat**, or home. Animals that live in **temperate**, or mild, climates may hibernate for only a few weeks. Garter snakes in the Arctic **tundra**, however, hibernate for about eight months each year! In every case, an animal's body changes to help it survive the harsh weather and lack of food.

 # Why do animals sleep?

Even if they do not hibernate, all animals rest or sleep. Sleep helps restore **energy**. Energy is the power animals need to do things. Animals use energy to breathe, grow, run, climb, fly, feed, and keep their bodies warm. If animals were awake all the time, they would quickly run out of energy and their bodies would grow weak.

Many animals cannot see well in the dark, so they sleep at night. When the day begins, they are well rested and can search for food. Other animals are able to see well at night. They sleep during the day and hunt from dusk to dawn.

Big cats, such as the leopard above, use a lot of energy when they hunt. They need plenty of sleep.

On the lookout

An animal's sleeping habits also depend on its enemies. Powerful **predators**, or hunters, such as tigers, have no enemies. They can sleep for long periods of time without waking. Animals that are hunted, such as deer and giraffes, can sleep for only short periods of time. They must be alert if predators approach.

How is hibernation different?

Animals sleep every day to save energy, but they must wake up and feed to get more energy. An animal's body **digests**, or breaks down, food and turns it into energy that the animal can use. Animals must also wake up so their bodies can get rid of liquid and solid waste.

Most animals would run out of energy if they slept nonstop and did not wake up to eat, but a hibernator can go months at a time without food. It uses almost no energy, so it is able to live off its stored body fat. Since the animal is not eating or drinking, it does not need to release waste. (See page 13.)

Most adults sleep about eight hours each night, but children need more sleep. How long do you sleep each night?

Giraffes do not get much sleep. They need to watch for predators, especially when they have offspring!

Feeding time

Hibernators must eat large amounts of food before their long sleep. They eat to gain body fat. During the late summer and early autumn, these animals eat all day. At this time of year, trees and bushes are full of nuts and berries. Some animals also collect food and store it to eat in the winter.

Both the ground squirrel, shown above, and the dormouse, on the left, are true hibernators. They eat all they can before winter.

Keep on eating!

Bears, mice, and squirrels prepare for winter by feeding almost constantly. Their bodies release special chemicals called **hormones**. Hormones make them want to eat and eat so that they will gain enough fat to survive winter. Some animals, such as bats, eat so much that they double their weight! Many animals also store extra food on which they can snack during winter.

Two types of fat

There are two types of fat found in hibernators—**white fat** and **brown fat**. The animal lives off white fat while it is hibernating. This fat burns slowly and lasts for several months. Brown fat is like rocket fuel. It is found near the heart and lungs because these organs are needed most for survival. When a hibernating animal is ready to wake up and needs a burst of energy, its body uses this quick-burning fat.

Raccoons fatten up before winter, but they are not true hibernators. They wake up and eat from time to time.

A perfect spot

Before an animal can hibernate, it must find and prepare a sleeping area. The spot must protect the animal from weather and hungry predators. Some animals use natural hollow spaces such as caves for their winter sleep. Other animals such as this groundhog use their sharp claws to dig holes in the ground. The weather outside may change daily, but the temperature underground changes very slowly. By sleeping in underground burrows, hibernating animals are safe from weather such as wind, rain, and snow. The soil acts as a wall that keeps out cold and holds in warmth.

Just big enough

An animal usually fits snugly inside its hibernation burrow. The cosy fit keeps the animal warm, just as a blanket keeps you warm when you sleep. If the burrow were too big, cold air could get inside it, and the animal would freeze. Some animals, such as the hoary marmot on page 10, line their dens with leaves, grasses, and twigs to make their sleeping places soft and comfortable. The lining also helps trap heat inside the dens.

*If an animal's body loses too much moisture, it **dehydrates**, or dries up, and the animal dies. Bats stay moist by hibernating in damp caves. Sleeping bats are often covered by water droplets.*

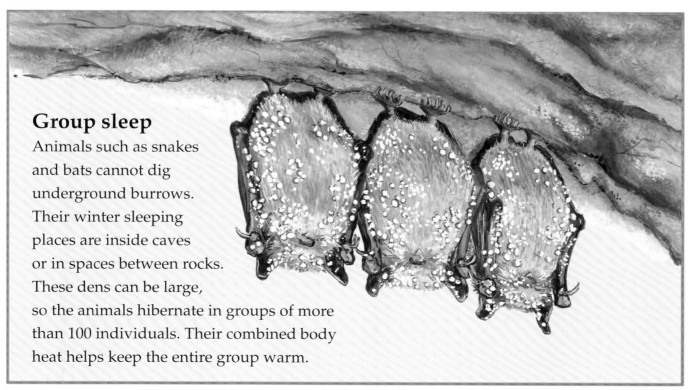

Group sleep

Animals such as snakes and bats cannot dig underground burrows. Their winter sleeping places are inside caves or in spaces between rocks. These dens can be large, so the animals hibernate in groups of more than 100 individuals. Their combined body heat helps keep the entire group warm.

Do not disturb!

Groundhogs are famous hibernators. The whole world watches to see if they can see their shadows when they emerge from their burrows in spring!

True hibernators go into the deepest winter sleep. They appear to be dead. You could pick up one of these animals without waking it! True hibernators are small **mammals** such as dormice, ground squirrels, and bats. Mammals are **warm-blooded**, which means their bodies usually stay the same warm temperature. While hibernating, however, true hibernators are able to lower their body temperatures to just above freezing in order to save energy.

Low and slow

True hibernators have the lowest body temperatures and slowest heart rates of all the winter sleepers. For example, as soon as a ground squirrel begins hibernating, its body temperature drops from around 100°F (38°C) to as low as 35°F (1.7°C). Its heart rate can fall from over 100 beats per minute to fewer than one beat per minute. The animal's heart still pumps blood through the body, but the blood moves at a very slow rate. It moves just fast enough to keep the animal alive.

Almost dead

When hibernating, a true hibernator breathes very slowly. Its heart barely beats, and its brain is almost totally inactive. True hibernation is different from sleep. During sleep, an animal's brain works constantly.

Tiny animals can

True hibernation occurs when an animal keeps a steady, low body temperature. Small mammals can be true hibernators because they can get cold and stay cold. Large animals cannot maintain a low enough body temperature to be true hibernators.

Food and water

Food is important to animals because of the **nutrients** it contains, but water is just as important. True hibernators cannot wake up to take a drink of water. Instead, they get water from their body fat.

No need to go

When an animal's body burns fat, it produces water. True hibernators keep this water in their bodies by not urinating. Normally, animals must urinate to release **urea**, a poison made after food is digested. Burning body fat does not produce urea, so hibernating animals do not need to urinate.

No frozen toes

A true hibernator such as this dormouse keeps its paws from freezing by curling them close to its body during hibernation. An animal's trunk is warmer than its arms and legs. If the animal keeps them drawn in close, its heart does not have to work as hard to warm the distant parts of the body.

Light sleepers

True hibernators are not the only mammals that sleep to survive winter. Many larger mammals, such as bears, skunks, and raccoons, use a lighter form of hibernation throughout these cold months. They can wake up on warmer days to watch for enemies, snack on stored food, or just have a stretch. To save energy while they hibernate, these animals keep their body temperatures much lower than normal. Their temperatures do not drop as low as the body temperatures of true hibernators, however. For example, a bear's temperature falls only a few degrees from 100°F (38°C) to around 93°F (34°C). The bear is too large to be a true hibernator because its body temperature cannot drop low enough and stay cold enough to save energy.

Above ground

Many lighter sleepers do not hibernate underground. Tree squirrels and raccoons often sleep in small groups inside hollow trees. They use the trees for nesting and raising their young year-round.

"I need more food!"

Like true hibernators, light sleepers such as squirrels live off body fat while hibernating, but they also eat food. On mild winter days, they wake up to eat some of the nuts they stored in autumn. This extra food replaces some of the fat they have lost.

Bears can wait

The nuts eaten by squirrels and other light sleepers contain **protein**. When the body burns protein, it produces urea, so these animals must urinate from time to time. A bear's body, however, is able to recycle its waste during winter. Like true hibernators, bears do not urinate.

Squirrels find the nuts they stored and snack on them during winter.

Beating the cold

Reptiles and amphibians are **cold-blooded**. The body temperatures of cold-blooded animals change as their surroundings get warmer or cooler. Cold-blooded animals do not make body heat, as warm-blooded animals do. They warm themselves by lying in the sun. Winter can be deadly for these animals, but many have found ways to beat the cold by hibernating.

In autumn, the days become shorter and cooler. Garter snakes, shown above, move underground to escape the cold air. A hibernating snake could not survive winter on its own—its blood would freeze. Instead, hundreds of snakes huddle together under the ground. The huge group creates a barrier against cold air, keeping most of the snakes warm enough to survive.

In the mud

Cold-blooded animals such as frogs and turtles hibernate underground at the bottom of ponds or streams. They burrow into the soft mud and wait for spring to warm the water again. Even though the surface of the pond freezes, the water near the bottom does not. The layers of ice, water, and mud help shelter the animals from the cold air outside. The animals breathe air bubbles that are trapped in the mud around them. They do not need much oxygen to survive because they breathe more slowly than they normally would.

Although frogs are cold-blooded like snakes, they do not need to hibernate in large groups for protection. They can dig a tight-fitting burrow that is as snug as a warm winter coat.

 # Snug as bugs

Insect **life cycles** are among the shortest on Earth. Insects hatch, become adults, and make babies faster than most other animals—sometimes in only a few days! Many insects, including butterflies and wasps, do not finish their life cycles before winter arrives. These insects must hibernate to survive until spring. They enter a type of hibernation called **diapause**. The growth of their bodies is "put on hold" until spring.

Most insects have three stages in their life cycles. Each starts out as an **egg**. After hatching, the young insect is called a **larva**. Eventually, the larva grows into an **adult**. When in diapause, insects do not grow into the next stage until spring arrives. Even insect eggs, such as the wasp eggs shown above, can go into diapause for the winter. They hatch in spring. The larvas then grow into young adult wasps.

Antifreeze

Insects are cold-blooded, and their tiny bodies store little heat. Insects such as wasps huddle in large groups, but unlike snakes, their bodies are too small to create a "blanket" against the cold. The body temperature of each still falls below freezing. Amazingly, their blood does not freeze! Insects have a type of **antifreeze** in their bodies. Antifreeze prevents liquids from freezing. It allows insects to survive until spring.

Many hibernating insects, such as these ladybugs, use houses and other buildings as shelter during winter.

Time to rest!

Monarch butterflies combine hibernation and migration to survive winter. In autumn, they leave Canada and the northern United States to fly thousands of miles to California and Mexico. They spend the winter hibernating on trees. In spring, they begin flying north. Along the way they stop, lay eggs, and die. Their babies hatch, and when they are fully grown, they continue the trip.

Water beds

When food grows scarce or water becomes too cold, most fish migrate to new areas, but a few types hibernate. Some nestle against the bottom of the lake or ocean, and others actually bury themselves in mud. Many fish, such as the mackerel above, eat tiny animals called **plankton**. During warm months, plankton live in large groups on the

water's surface, where it is sunny. As the weather cools, the plankton begin to disappear, and the mackerel cannot find enough food. The fish swim to deeper, cooler waters. Their breathing slows down, and they lie motionless on the ocean floor. In spring, plankton return to the sunny surface and so do the mackerel that feed on them.

Swimming deeper

The surfaces of lakes and ponds often freeze in winter. Fish escape the ice by swimming into deeper water. Most fish do not eat much during this cold time. They stop growing for a while, but they do not actually hibernate.

Rolling in the mud

The carp is one of the few lake fish that do hibernate. When a carp reaches deeper water, it heads for the bottom. It thrashes around in the mud and eventually buries itself. Like other hibernators, it breathes only half as often as it normally would.

Carp are well known for their "thrashing" behavior. A carp thrashes to dig a hole for itself before it hibernates. During the rest of the year, it thrashes to find food on the lake bottom.

Tweet dreams

As winter approaches, most birds, including the terns shown above, migrate to warmer places. For a long time, scientists did not believe that any birds hibernated. Then they learned that poorwills hibernate throughout the entire winter! These birds sleep inside hollow logs or in natural holes between rocks. When a poorwill is hibernating, its breathing slows down, and its body temperature drops from 100°F (38°C) to 65°F (18°C).

This hibernating poorwill is well camouflaged so that predators cannot find it while it sleeps.

Restful nights

The tiny hummingbird can flap its wings faster than any other bird can. This fast action uses a lot of energy. When a hummingbird sleeps each night, its rest resembles hibernation. During sleep, the bird does not move, and its body temperature drops. Although the hummingbird "hibernates" only overnight, the deep sleep saves its body valuable energy.

A hummingbird also sleeps to save energy during long periods of bad weather, when it is difficult to find food.

What is estivation?

Animals that live in cold places are not the only hibernators. Animals in hot areas also go into a deep sleep. This type of hibernation is called **estivation**. These animals estivate to survive **dry seasons**, when very little rain falls and there is not much food to eat. They sleep underground for several months to escape the heat and to save body moisture until the rains return.

Like hibernating frogs and turtles, estivating alligators and crocodiles bury themselves in mud at the bottom of streams. Instead of keeping them warm, however, the mud keeps these animals cool and moist. The large reptiles have no natural enemies, so they can lie patiently in one place. There are no predators to sneak up and attack them.

Desert tortoises estivate in the hottest months by burying themselves in sand.

Desert life

Deserts are dry places that get very little rain. The tiny reptiles and mammals living there have adjusted to life with almost no water. In fact, desert animals such as skinks and jerboas get most of the moisture they need from the insects they eat. To survive the hottest months, however, these animals dig burrows deep under the sand, where they estivate. The burrows help protect the animals from heat and predators.

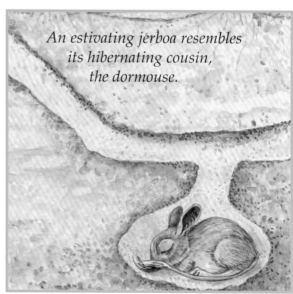

An estivating jerboa resembles its hibernating cousin, the dormouse.

Waiting for rain

A long period without rain is called a **drought**. Some droughts last a few years. Even an estivating animal cannot survive without water for that long! A few animals, however, can become **dormant**, or inactive, and survive for years without water or food.

Snails have soft bodies that must remain moist. During droughts, a snail pulls its whole body inside its shell. It then makes a hard layer that seals the opening of the shell. This "door" helps the snail trap moisture. When it rains, the snail comes out of its shell again.

snails

Tiny survivors

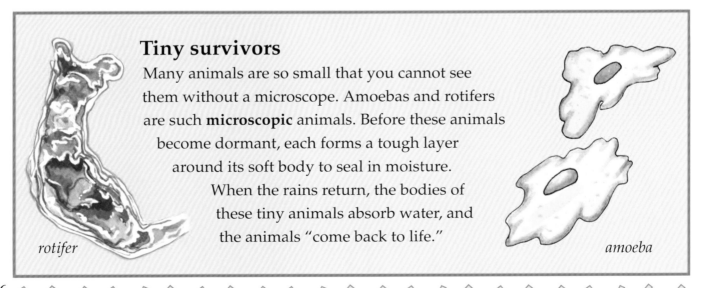

Many animals are so small that you cannot see them without a microscope. Amoebas and rotifers are such **microscopic** animals. Before these animals become dormant, each forms a tough layer around its soft body to seal in moisture. When the rains return, the bodies of these tiny animals absorb water, and the animals "come back to life."

rotifer

amoeba

Lungfish

Fish use **gills** to breathe. Most fish cannot breathe out of water, but a lungfish is able to breathe both in and out of water. Lungfish live in rivers and shallow ponds in Africa and South America.

When these ponds and rivers dry up, the lungfish burrows into the mud at the bottom. It creates a slimy cocoon around its body to help seal in moisture. The fish also leaves holes in the mud so it can have air to breathe while it is in a dormant state.

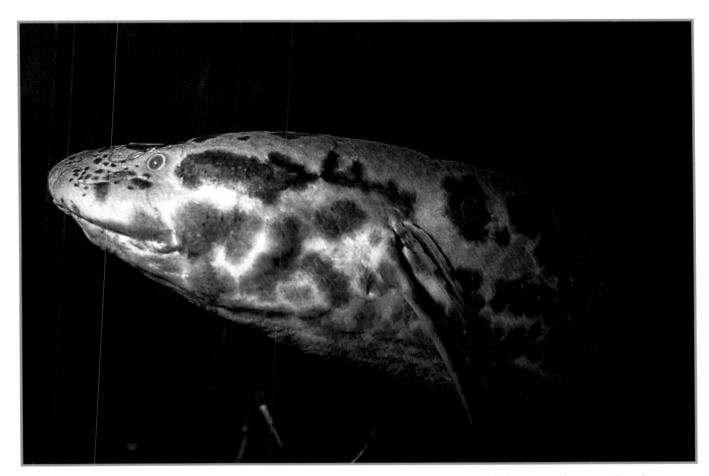

The lungfish can remain dormant for as long as twenty years. Some amphibians, such as toads, also become dormant to survive droughts.

27

The nursery

Young animals are always at risk. They have difficulty finding food and are easy targets for predators. Some animal mothers use their hibernation time to give birth and raise their young.

A female bear, or **sow**, gives birth to cubs in her den. The newborn cubs are blind, hairless, and extremely tiny—you could hold one in the palm of your hand!

The sow's body turns her fat into milk for the cubs to drink, and produces heat to help keep the den warm. The cubs spend much of the winter awake, even though their mother is asleep. They curl up close to her body to rest when they are tired. Cubs are safe from predators in their snug den. They can grow bigger and gain strength before spring arrives.

A mother bear provides food and warmth for her cubs. She can also wake up quickly to protect her young if an enemy comes near.

Frozen eggs

Many animals are experts at protecting their young from cold or drought. Desert toad eggs, for example, can become dormant. If a pool dries up, the toads lay eggs covered in slimy cocoons. Even if the parents do not survive, the young tadpoles will hatch when the rains return. Many insects lay their eggs underground or inside logs in autumn. The eggs go into diapause and hatch in spring. (See page 18.)

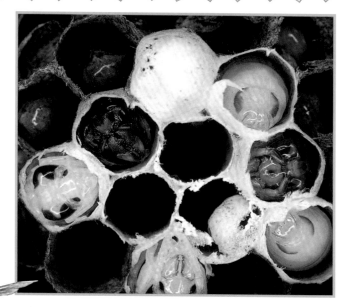

These wasp eggs will not hatch until the weather is warm enough for the baby wasps to survive.

Safety in numbers

Garter snakes give birth just before they hibernate. The youngsters follow their parents underground. For warmth, they coil near the middle of the huge group of sleeping snakes so they will not freeze. (See page 16.)

Kangaroo babies on hold

Female kangaroos can give birth only when water is available in their dry habitat. Their bodies need extra water to make milk for their **joeys**, or babies. After two kangaroos mate, the female stores her eggs inside her body. An egg does not **gestate**, or continue to grow, until there is enough water to make milk.

I'm awake!

Once spring returns, hibernating animals must wake up. Some animals, such as raccoons and bears, wake up often during hibernation to stretch and move around. Others, such as dormice, bats, snakes, and insects, are nearly frozen. They cannot move when they wake up. They must wait for their bodies to warm up before they can move out of their dens.

Scientists are not sure how some animals know when to wake up. Light sleepers such as bears leave their dens often enough to notice the weather changing, but how do true hibernators know when it is time to awaken? As spring approaches, the days grow longer. Some scientists believe that once the days reach a certain length, the animals know that their winter sleep is over.

Cold-blooded animals, such as this leopard frog, bask in the sun to warm up after hibernation.

Burning fat

True hibernators burn their stored brown fat to wake up in spring. This fat warms up the heart, brain, and lungs. The animal breathes more quickly, and its heartbeat speeds up. Its heart pumps blood into its legs and tail. The animal is now almost ready to leave its den.

Ground squirrels eat the last of the food they stored during autumn, but many animals keep living off their body fat until they find something to eat. Large animals such as bears find food as soon as they leave their dens. They teach their cubs how to catch tasty fish such as salmon.

Glossary

amoeba A tiny one-celled creature

burrow (n) An animal's underground home; (v) to dig underground

dehydrate To lose moisture

diapause A type of hibernation during which the eggs or bodies of some animals and insects stop growing

dormant Describing something that is not in an active state

drought A long period without rain

gestate To grow inside a mother's body

ground squirrel Referring to the Arctic ground squirrel, a rodent that lives in burrows and is a true hibernator

hormone A chemical produced by an animal that helps regulate its body

larva A young insect after it has hatched

life cycle A set of changes from the time an animal is born to the time it is an adult that is able to make babies

microscopic Describing the size of animals that can be seen only with a microscope

nutrient A natural substance that helps animals and plants grow

predator An animal that hunts and eats other animals

protein A substance needed for growth that is found in foods such as meat

rotifer A tiny creature with a body that is made of a few cells

temperate Describing mild weather

tundra A treeless plain in the Arctic

Index